SCIENCE FIELD TRIPS
Inside, Outside, and Off Campus

by Pam Walker and Elaine Wood

illustrated by Janet Armbrust

cover by Linda Pierce

Publisher
Instructional Fair • TS Denison
Grand Rapids, Michigan 49544

ISBN: 1-56822-846-5
Science Field Trips
Copyright © 1999 by Ideal · Instructional Fair
a division of Tribune Education
2400 Turner Avenue NW
Grand Rapids, Michigan 49544

TABLE OF CONTENTS

TO THE TEACHER

Science is everywhere, and the best way to learn it is to experience it. That is why taking students out of the classroom can be an exciting and rewarding experience. Nothing makes a concept as real as learning to apply it to everyday life. And to do that, teachers must stretch their legs and get their students out of their room.

This book is designed to make it easy for the teacher to get up and out. Each field trip lesson provides the student with some background information and one or more student activity pages to be completed during the trip. This strategy helps students focus their attention on the topic you are teaching and keeps them actively engaged in learning.

On the teacher information page of each activity, there is a grading rubric to help evaluate students' work. Little or no equipment is needed for the activities.

There are several kinds of field trips that teachers and students can take.

1. Field Trips in the School Building
   How often do you explain to your students that the techniques of measurement they learn in class are required skills in many jobs? Do you remind them that examples of the simple machines they study in physical science can be found all over the school building? Have they surveyed their school environment to determine how dependent we all are on wood products?

   Take students out of the classroom and into other parts of the school. Have them really examine the components of their building.

2. Field Trips on the School Campus
   Students are somewhat interested in their book's description of a worm or a seed. However, they are excited at the prospect of actually examining the real thing. Schools are not isolated from the real world, and by just walking out the school doors, you can connect your students to real-life science.

3. Field Trips off Campus
   Load everyone into a school bus and take students to see the world. Instead of talking about endangered animals, visit them. Do not try to explain how stalactites look; go see some. Nothing impresses students more than a trip and a change in their routine. Take advantage of the facilities near your school, whether they are tourist attractions like zoos or public utilities like sewage treatment plants.

Inside

Text in image: TISSUE, WOOD, WOOD PROJECTS, THINGS MADE FROM WOOD

THE WONDERS OF WOOD

Teacher Information

TIME REQUIRED

One to two hours, depending on the number of places visited

Objectives: Students will identify ways that products from the forest are used inside the school building.

Teaching Strategies: Before this trip, discuss some of the products the forest gives us. Remind your class that trees yield products made from paper (books, cards, wrapping paper), cellulose (carpet, Ping-Pong balls, luggage), bark (bottle corks), and resins (cosmetics, rubber gloves).

Divide the class into small groups of two or three. Each group should select a person to serve as the data recorder who carries the student activity page. During the tour, students should observe their surroundings closely.

Instruct students to read the Background Information and answer the Pre-Lab Questions.

You may wish to have students tour locations, such as the auditorium, principal's office, cafeteria, gym, bathroom, special classrooms, and the library.

Evaluation: At the conclusion of the field trip, each group should turn in the completed student activity pages.

A suggested grading rubric:

Criteria	Points allowed	Points awarded
Pre-Lab Questions correct	20	_____
Group was on task during the trip	20	_____
Product Chart completed	20	_____
Products ranked	10	_____
Thought questions answered	30	_____
Total	100	_____

THE WONDERS OF WOOD

What good is the forest to you and me? Forests provide us with a variety of material that we use every day, such as food, rubber, paper, wood, and medicine. In addition, forests play a larger role in the ecosystem. They are vitally important in producing oxygen, preventing erosion, and moderating the climate.

Have you ever thought about all the uses for a tree? You may have seen tall pines being cut down. Sometimes, workers use chain saws to remove the limbs from the tree and cut the tree into more manageable pieces. Depending on their intended destinations, these pieces are loaded onto trucks and hauled away. On very large tree farms, specialized equipment cuts the trees, strips their limbs and bark, and then loads them on a truck in one efficient process.

Figure 1. Trees are cut, stripped, and loaded with the same piece of equipment.

There are four basic categories of products that are made from trees:

A. Paper products—used in objects, such as books, cereal boxes, wrapping paper, facial tissue, toilet paper, and newspapers.

B. Cellulose products—used in carpet, pillows, Ping-Pong balls, shampoo, plastic film, insulation, toothbrush handles, and wallpaper paste.

C. Bark products—used in corks for wine bottles, the cork center of a baseball, and many types of medicines.

D. Resin products—used in cosmetics, paint thinner, coating for pills, rubber gloves, and paint.

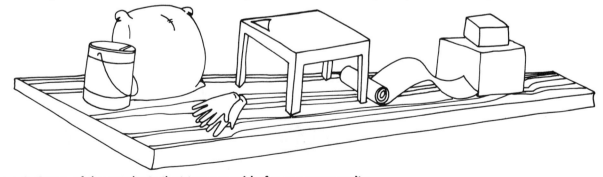

Figure 2. Some of the products that trees provide for our community

These categories do not even include the foods that trees produce. Trees give us a variety of fruits and nuts, such as pecans, walnuts, limes, dates, cherries, and cloves. The United States forest products industry provides jobs for millions of people from foresters to mill workers to engineers.

Pre-Lab Questions

1. List some products made from trees. _____
_____
_____

2. In the ecological sense, why are trees important?_____
_____
_____

3. Recycling tree products saves money and consumes fewer trees. Name one tree product that can be recycled. _____

4. You are a pine tree standing in the forest. Workers arrive to cut you down and change you into lumber that can be used for construction. Describe everything that happens to you from the time you are cut until you become a piece of lumber._____
_____
_____
_____

THE WONDERS OF WOOD

I. Data Collection—In the table below list all of the forest products you see while on your tour. Record the locations where you see each product and group the products into one of four categories: Paper, Cellulose, Bark, or Resins. Do not fill in the ranking column at this time. The first example is done for you.

Name of product	Location in which product was found	Group in which the product belongs	Ranking of importance to your group
Example—Facial tissue	Main office—on secretary's desk	Cellulose	

II. The Ranking—After your tour, go back and look at the products you listed that came from trees. Think about the importance of each of these products to society. On the chart, rank each product you listed from 1 to 23, with 1 being the most important and 23 being the least important.

III. Thought Questions—Answer the following questions in detail.
   1. For each product with a ranking over 10, suggest a material that you could use as an alternative to the material derived from trees. Explain why you chose the alternative.

   _____

   _____

   _____

   _____

   _____

   _____

   2. If all the trees were eliminated from earth, how would this change life on earth?

   _____

   _____

   _____

MEASUREMENT MANIA

TIME REQUIRED

One to two hours

Objectives: Students will use their knowledge of the metric system to calculate the volume and area of three hallways in their school.

Teaching Strategies: Prior to this field trip, students should read the Background Information and answer the Pre-Lab Questions. You may want to have students practice measuring things with meter sticks. Work some sample volume and area problems on the board. Explain that area is calculated in square meters and volume in cubic meters.

Divide the class into small groups of two or three. Each group needs the student activity pages, a string, and a meter stick. One student in each group should serve as data recorder.

Instruct students to calculate the length, width, and height of the hallway they are assigned. Tell students to think of the hallway as a structure that is completely enclosed on all sides. Their data for each of these dimensions should be recorded in meters.

Evaluation: At the conclusion of the field trip, each group should turn in its completed student activity pages.

A suggested grading rubric:

Criteria	Points allowed	Points awarded
Pre-Lab Questions correct	20	_____
Group was on task during the activity	20	_____
Measurements were done accurately	30	_____
Student activity pages correct	30	_____
Total points	100	_____

MEASUREMENT MANIA

The last time you were in the grocery store, did you notice that the weight or volume of many products is given in English and metric units? For example, the quantity of sugar in a bag is reported as one pound (1 lb) and as 454 grams (454 g). You are familiar with pounds as a unit of measurement. The unit of grams may be new to you, but it is part of a system of measurement that is recognized all over the world. Grams is one of the units of the metric system. Meters and liters are also metric units.

Metric units are based on the powers of 10. A meter is the basic unit for distance in the metric system. A meter is a little longer than a yard. Units that are smaller and larger divisions of a meter are written using prefixes:

Milli	-	$\frac{1}{1,000}$
Centi	-	$\frac{1}{100}$
Deci	-	$\frac{1}{10}$
Meter	-	1
Deka	-	10
Hecta	-	100
Kilo		1,000

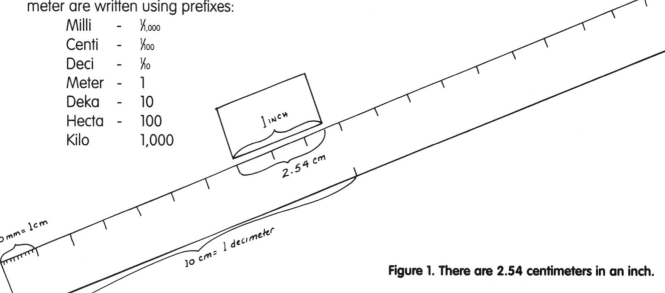

Figure 1. There are 2.54 centimeters in an inch.

The same prefixes such as milli-, centi-, and kilo- are used with units of mass, called **grams,** and units of volume, called **liters**.

People use measurement every day for a variety of reasons. Carpenters and architects often need to know the area of a certain location. Some products, such as flooring, are sold by area. Area can be determined by multiplying length times width.

area = length x width

How could you calculate the area occupied by a small building? You could measure the length of the building in meters and then the width of the building in meters. Then multiply these two numbers to find the area in square meters.

Another measurement that we often need to know is volume. Volume refers to the amount of space occupied by something. For example, you can calculate the volume of a container or a building by multiplying length times width times height.

volume = length x width x height

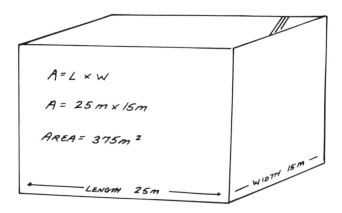

Figure 2. A building that is 25 meters in length and 15 meters in width would occupy an area of 375 square meters.

To determine the volume of space your classroom occupies, you would measure the length, width, and height of your room. If these measurements are recorded in meters, the answer would be expressed in cubic meters.

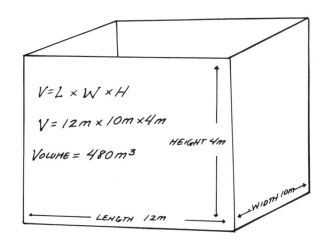

Figure 3. If your classroom is 12 meters long, 10 meters wide, and 4 meters high, it has a volume of 480 cubic meters.

Pre-Lab Questions

1. What is the metric unit of length? _____ mass? _____

2. How can area be calculated? _____

3. If the two dimensions of an area were calculated in meters, in what units would you report the area? _____

4. How can volume be calculated? _____

5. If the three dimensions of a volume were calculated in meters, in what units would you report the volume? _____

MEASUREMENT MANIA

I. Measurement

Use your string and your meter stick to measure the length, width, and height of each hallway. Record this information in Data Table I. Log all your measurements in meters, recording your answer two places past the decimal.

Data Table I

	Hallway 1	Hallway 2	Hallway 3
Length			
Width			
Height			

II. Calculations

Use the information you recorded in Data Table I to calculate the area and the volume of each of the three hallways. Show your work in the appropriate space below. Record your answers two places past the decimal and place the proper units beside your numerical entries. Write your answers in Data Table II.

Data Table II

	Hallway 1	Hallway 2	Hallway 3
Area			
Volume			

Conclusion Questions

1. Did these three hallways have the same area? _____ the same volume? _____

2. If you did not have a string and a meter stick, but knew the length of your foot in meters, how could you have calculated the length and width of the hall?

   _____

   _____

3. At one time the unit length was based on the length of the foot of a famous king. Why would this not be an acceptable unit of measurement today?

   _____

   _____

4. Explain why it is important to use standard units of measurement in science.

   _____

   _____

MACHINE MANEUVERS

TIME REQUIRED

One to two hours, depending on the number of places visited

Objectives: Students will locate and list examples of the six types of simple machines in the school building.

Teaching Strategies: Before leaving the classroom, students should read the Background Information and answer the Pre-Lab Questions. Discuss some common examples of each type of simple machine and explain the three classes of levers.

Divide the class into groups of two or three. Each group should select a recorder who carries the student activity pages and records data for the group.

Instruct students to observe things closely as they tour the building. During the tour, the recorder should write down all of the simple machines observed by the group. Each group should classify the machines they locate by type on the student activity pages. In the case of levers, the group should indicate whether they are first, second, or third class.

You may wish your tour to include locations such as the auditorium, principal's office, cafeteria, gym, bathroom, and special classrooms.

Evaluation: At the end of the field trip, each group should turn in completed student activity pages.

A suggested grading rubric:

Criteria	Points allowed	Points awarded
Pre-Lab Questions completed	20	_____
Group was on task during trip	20	_____
At least 10 machines correctly identified in Data Table I	20	_____
At least 10 machines correctly identified in Data Table II	20	_____
Each machine was classified correctly	20	_____
Total	100	_____

MACHINE MANEUVERS

Have you ever watched a professional mover load a large object, like a piano, onto a truck? Did he pick up the piano and set it inside the truck? If so, he was incredibly strong. It is more likely that he placed the piano on rollers and pushed it up a ramp to the truck. A ramp is a type of simple machine called an **inclined plane**. Simple machines make work easier for us.

There are six basic types of simple machines. All are useful because they reduce the amount of work someone must do. Work is equal to force exerted by a person over a distance. Simple machines allow people to do jobs that they ordinarily would not be strong enough to perform, usually by reducing the amount of force required.

The six types of simple machines include the following:
  a. A lever is a bar that is used to pry or dislodge an object, such as a screwdriver that is placed under the cap on a paint can to pry it loose. All levers have three common features: a fulcrum (pivot point), an effort arm, and a resistance arm.

There are three kinds of levers.
  1. A **first-class lever** has the fulcrum positioned between the effort force and the resistance force, like a seesaw or a pair of scissors.

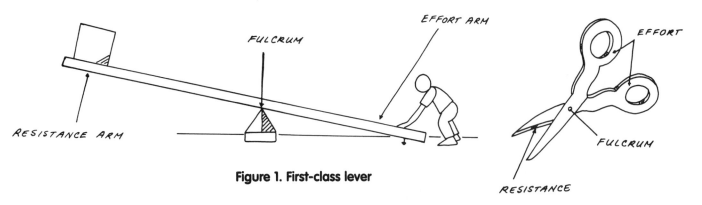

Figure 1. First-class lever

  2. A **second-class lever** has the resistance force located between the fulcrum and the effort force, like a wheelbarrow.

Figure 2. Second-class lever

3. A **third-class lever** has the effort force located between the fulcrum and the resistance force. When a tennis player hits a ball, his/her arm is a third-class lever.

Figure 3. Third-class lever

b. A **wheel and axle** is really a lever connected to a shaft or rod. The wheel is fixed on the axle and spins on it. A steering wheel of an automobile is an example of a wheel and axle.

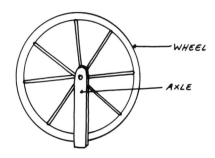

Figure 4. Wheel and axle

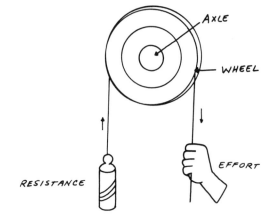

Figure 5. A pulley

c. A **pulley** is made from a wheel that is free to spin on an axle. A block and tackle, such as the type used to lower a large piece of furniture out of a window, is an example of a pulley.

d. An **inclined plane** is a flat slope on which objects can be rolled or pushed upward. A ramp leading from the ground to the edge of a moving truck is an inclined plane.

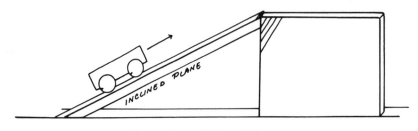

Figure 6. An inclined plane

e. A **wedge** is a flat object that is narrow on one end and wide on the other. It is used to separate objects. An ax is a wedge.

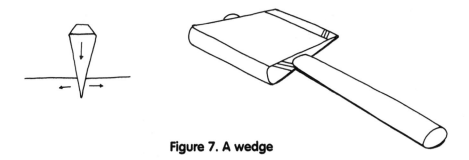

Figure 7. A wedge

f. A **screw** is a type of inclined plane that is wrapped around a cylinder. Screws are used to attach objects to each other.

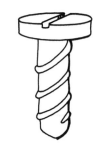

Figure 8. A screw

Pre-Lab Questions

1. What are the six types of simple machines? _____

_____

_____

2. Why do we use simple machines? _____

_____

3. How does an inclined plane reduce the amount of force required by a person to move a large object? _____

_____

MACHINE MANEUVERS

I. In Data Table I, list the simple machines that you see on your tour of the school. Give their locations and classify them as lever, pulley, wheel and axle, inclined plane, wedge, or screw. If the machine is a lever, label it as a first-, second-, or third-class lever. Two examples have been done for you.

Data Table I

Name of object	Location	Type of machine	If lever, what class?
Example—Custodian sweeping floor	Hallway outside bathroom	Lever	Third
Example—A pencil sharpener	Library	Wheel and axle	

II. With the members of your group, list ten simple machines in Data Table II that could be found outside the building on your campus.

Data Table II

Example of machine	Category of machines	Illustration of machine
Example—Steering wheel in the principal's car	Wheel and axle	

MAKING SENSE OF A SITUATION

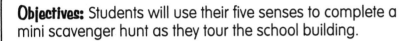

TIME REQUIRED
One to two hours, depending on length of time spent at each location

Objectives: Students will use their five senses to complete a mini scavenger hunt as they tour the school building.

Teaching Strategies: Before the trip, discuss the five senses and how important each is to our survival. Ask students what senses they are using at this time. On the board, list the current uses of each sense.

Our senses give us information about our environment. Point out that dogs and many other animals have a sense of smell that is hundreds of times more sensitive than ours. Ask students to imagine how they might perceive the room they are in if they could smell as well as a dog.

Have students imagine how they might react to situations if they lost one of their senses. Find out if they know anyone who has lost a sense. What problems does that person have coping with daily life?

Instruct students to read the Background Information and answer the Pre-Lab Questions.

Some locations that you could visit on your tour of the school include the gym, library, office, auditorium, chorus room, and art room. Have students complete the student activity pages as they are touring the school.

Evaluation: At the end of the field trip, each student should turn in a completed student activity page.

A suggested grading rubric:

Criteria	Points allowed	Points awarded
Pre-Lab Questions completed	20	_____
Group was on task during trip	20	_____
Student activity pages completed	40	_____
Follow-up activity completed	20	_____
Total	100	_____

MAKING SENSE OF A SITUATION

How do you know what is happening in your environment? As humans, we get most of our information about the environment by sight. But we get additional information from four other senses: hearing, touch, smell, and taste. Together, these five senses keep us informed about changes in our surroundings.

Light enters the eye through the transparent cornea. The amount of light entering the eye is controlled by the iris. As light passes through the lens, it is bent and focused on the retina at the back of the eye. These receptors are connected to nerves that send information about light perception to the brain.

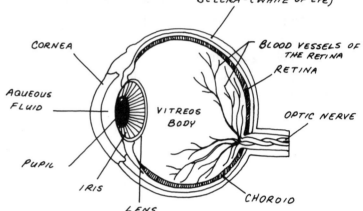

Figure 1. Parts of the eye

The ear is a complicated sense organ inside the head. The part of the ear that we can see simply helps collect sound waves. When a sound wave enters the ear canal, it causes the eardrum to vibrate. This in turn causes three small bones attached to it to vibrate. These bones vibrate a fluid-filled chamber that in turn pushes against sensory hairs. Bending the hairs causes them to send nerve impulses to the brain.

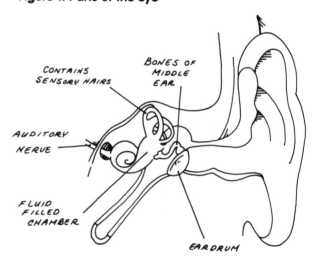

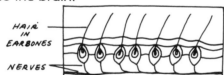

Figure 2. Parts of the ear

The senses of taste and smell are both chemical senses. Embedded in the tongue are thousands of receptors called **taste buds**. These receptors are specialized to detect certain kinds of chemicals: sweet, sour, bitter, and salty. Generally, the tip of the tongue is most sensitive to sweet, the sides to sour, and the back to bitter. The smell receptors are a patch of cells in the top of the nasal passage. There are about 5 million cells in this area that can detect certain chemicals in the gaseous state. The senses of taste and smell work closely together.

The sense of touch is due to nerve receptors located in the skin. Changes in pressure on the skin affect the shape of these receptors. When the receptors change in any way, they send signals to the brain which we interpret as touch.

Pre-Lab Questions

1. Why do animals need their five senses? _____

2. On which sense do humans depend the most? _____

3. Explain how we perceive things by the sense of sight. _____
   _____
   _____

4. Explain how we perceive things by the sense of smell. _____
   _____
   _____

5. How are the senses of taste and smell related? _____
   _____

6. How do we perceive things by the sense of touch? _____
   _____

7. Where do all of our senses send the information they detect? _____
   _____

MAKING SENSE OF A SITUATION

I. Completion—As you tour the school building, write down an example of something that fits one of the descriptions in the first column in the Data Table. In the second column, indicate where the item was located. After all of the examples are recorded, go back through the table and name the sense or senses that you used in the third column.

Data Table

Descriptions	Name of item that fits the description	Place item was located	Sense(s) used to detect the item
Something cold			
Something rough			
Something made from a tree part			
A form of waste material			
An insect			
A habitat for something			
An object that can aid the functioning of one of your senses			
Something made from a renewable natural resource			
Something that photosynthesizes			
A heterotroph			

Evidence of animal activity			
A form of pollution			
A compound			
A mixture			
A good insulator			
Something that generates heat			
Something cooler than the air			
Something that makes a loud noise			
Something with a density greater than water's			
A material that will crumble if you place enough pressure on it			
Something that generates light energy			
Something that needs water to function			
Something that provides us with energy			

II. Follow-Up Activity

Pretend that you were blindfolded on your tour today. In other words, you had no use of your sense of sight. Which of the 23 items that you identified today would have been impossible to locate without your sense of sight?

MORE POWER TO THE GAMES

TIME REQUIRED

One to two hours

Objectives: Students will calculate the amount of work and power involved in a three-legged race.

Teaching Strategies: Prior to this adventure, solve some problems on power and work on the board. One of the easiest ways to measure force in newtons is with a spring balance. Below are some sample problems.

1. You connect a block of wood to a spring balance and pull the wood .5 meters. The spring balance shows you that you are exerting a force of 2 newtons. How much work is being done?

    W = F x d   W = 2 newtons x .5 meters   W = 1.0 joules

2. You pull this same block of wood a distance of .5 meters up an inclined plane. This time, the spring balance shows that you are exerting .8 newtons of force. How much work is being done?

    W = F x d   W = .8 newtons x . 5 meters   W = .40 joules

3. It took four seconds to do the work that was performed in Problem # 1. How much power was used?

    P = W/t   P = 1.0 joules / 4 sec   P = .25 watts

Take students to an open area of the school, such as the gym or a hall. Mark off a "race course" that is 100 meters long. Divide the class into groups of three or four. Remind students to wear comfortable shoes for this activity.

This activity will require the following for each group:

Strips of rags to tie around the ankles of students   Stopwatches
Kilogram balances   Calculators
Meter stick or tape measure

Evaluation: At the conclusion of the field trip, each group should turn in the completed student activity pages.

A suggested grading rubric:

Criteria	Points allowed	Points awarded
Pre-Lab Questions correct	20	_____
Group was on task during the activity	30	_____
Proper recording of information in data table	20	_____
Student activity pages completed	30	_____
Total points	100	_____

MORE POWER TO THE GAMES

A boulder has rolled onto the road and is blocking it. You attempt to push the boulder out of the way, but it will not budge. After several hours of pushing and sweating, you give up. You feel that you really worked on it. Actually, you did no work at all. According to science, the definition of work is "the movement of a force over a distance." You were unable to move the object even though you used a lot of force. Since no distance was achieved, you did not do work.

Work can be calculated by multiplying force used times the distance through which the force moved.

$$W = F \times d$$

Force is measured in units called **newtons**. A newton can be calculated by multiplying mass in kilograms times acceleration in meters per second squared. Distance is measured in meters. Therefore, when you multiply a force (in newtons) times a distance (in meters), your answer has units of newton-meters. For convenience, a newton-meter is also called a **joule**. Therefore, joules are units of work.

You can calculate the force you generate if you know your mass. Your mass multiplied by 10 m/sec^2 (which is acceleration due to gravity) will give you the force in newtons.

Figure 1. You perform work when you jump 2 meters from the ground. First calculate the force your body exerts: if your mass is 50 kg, multiply 50 kg x 10 m/sec/sec. This calculation tells you that your body exerts a force of 500 newtons. Then multiply 500 newtons times 2 meters, and you find that by jumping you performed 1,000 joules of work.

If you had a stopwatch and timed how many seconds it took you to complete that jump, you could find the power you exerted. Power reflects work done over time. In our calculation, time should be measured in seconds. The unit of power is watts.

Power = Work/time

Figure 2. If it took you 2 seconds to perform that jump, you generated a power of 1,000 joules/2 seconds or 500 watts of power.

Pre-Lab Questions

1. How do you calculate the force your body exerts? _____
   _____

2. If you weigh 40 kilograms, how much force does your body exert when you move?
   _____

3. How do you calculate work? _____
   _____

4. How do you calculate power?_____
   _____

5. In which case was more power exerted: a horse that tried to pull a 2,000-kilogram car; a child who lifted a .5 kilogram book a distance of 1 meter? _____Why?
   _____
   _____
   _____

More Power to the Games

Select one person in your group to be the time keeper. The other two members will be the three-legged race participants. The time keeper should tie a scarf or rag around the two inside legs of the participants while they stand side by side, facing in the same direction.

At the direction of your teacher, each participating pair should assume their positions at the starting line. The time keepers will position themselves so that they can see both the starting and the ending line. When the teacher blows the whistle, the participants should walk as quickly as possible to the finish line. The time keepers will start the stopwatches at the whistle and stop them when the groups cross the finish line. The course is 100 meters long.

Figure 3. Two people tied together for three-legged race

In Data Table I calculate the work and the power generated by the participants from your group. Participants should find their mass using a kilogram balance. Add the two masses together to get a total mass.

Repeat the activity two more times, using a different combination of participants each time.

Data Table I

Work done by first pair of participants	Work done by second pair of participants	Work done by third pair of participants

Data Table II

Power produced by first pair of participants	Power produced by second pair of participants	Power produced by third pair of participants

Conclusion Questions

1. If a group ran two races, a 100-meter race and a 50-meter race, during which race was the most work done? (Assume that the mass of the group remained the same in both races.) _____ Why? _____

   _____

2. In which race did the group produce the most power: in the race that lasted three seconds or in the race that lasted five seconds? _____ Why? _____

   _____

3. You pushed for 30 seconds on a brick wall, but the wall did not budge. You have a mass of 50 kilograms. How much work did you do? _____ How much power was generated? _____

4. Where do you think the term **horsepower** originated? _____
   Explain your answer. _____

   _____

   _____

JOURNEY TO YELLOWSTONE

TIME REQUIRED

Two to three hours

Objectives: Students will utilize the Internet to take a "virtual field trip" to Yellowstone National Park.

Teaching Strategies: Prior to the trip demonstrate to students how to use the Internet and its various search engines.

Divide the class into groups of two or more. Give each group an activity sheet and instruct them to appoint one of the group members as the recorder of data.

Have students read the Background Information on the history of Yellowstone and explain terms they may encounter during their search. Have them answer the Pre-Lab Questions before they access the Internet.

Evaluation: At the conclusion of the field trip, each group should turn in the completed student activity pages.

A suggested grading rubric:

Criteria	Points allowed	Points awarded
Pre-Lab Questions correct	20	_____
Group was on task during the trip	20	_____
Student activity pages completed	30	_____
Travel brochure about Yellowstone completed	30	_____
Total	100	_____

JOURNEY TO YELLOWSTONE

Yellowstone National Park, located on 2 million acres in the northwest corner of Wyoming, was founded in 1872. The park is visited by more than 3 million people each year. It is composed of about 80% forests, 15% grassland, and 5% water. The highest point in the park is Eagle Peak at an elevation of 11,000 feet. Yellowstone Lake covers about 110 miles of park shoreline.

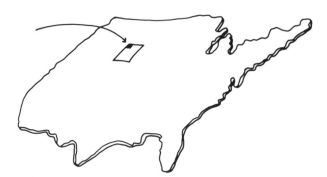

Figure 1. Map of the United States showing location of Yellowstone National Park

The park is home to over 250 active geysers. Geysers occur when water heated by the earth's interior reaches the surface and erupts. In active geysers, the heat causes spectacular streams of water and steam to shoot high into the air. In fact, there are more geysers in Yellowstone than anywhere else on earth.

Another spectacular site is the Grand Canyon of Yellowstone. This canyon is 1,200 feet deep and highlighted by the powerful Upper Falls. This canyon is still being cut by the Colorado River that runs through it.

Because Yellowstone is a refuge for all forms of wildlife, tourists usually see a wide variety of animals. As you travel through Yellowstone, you cannot miss the bison and, if you are lucky, you may see bears. Park officials urge visitors to avoid approaching or contacting these animals. Bison are about six feet tall, weigh over 2,000 pounds, and can run over 30 miles per hour. In fact, some visitors each year have been gored by the animals because they ventured too close.

Figure 2. Bison grazing in Yellowstone National Park

Visitors who feed animals are not doing them any favors. Feeding wild animals makes them dependent on people for their food. These animals do not learn to find food for themselves. Therefore, when visitors are not around with handouts, the animals go hungry.

While in the park several special rules apply. If you take a pet into the park, it must remain on a leash at all times and it may not walk on the trails. Campfires are permitted only in designated areas, and you are not allowed to pick wildflowers or to collect natural objects from the park.

Pre-Lab Questions

1. In what state is Yellowstone National Park located? _____

2. What animal has injured visitors who did not follow park instructions? _____

3. What are there more of in Yellowstone than anywhere else in the world? _____

4. How are geysers formed? _____
   _____

5. Have you or any member of your group ever visited Yellowstone? _____ If so, what can you remember about your trip? _____
   _____
   _____
   _____
   _____
   _____

6. Why is it not a good idea to feed the animals in national parks? _____
   _____
   _____

Name _____

JOURNEY TO YELLOWSTONE

You and your group are going on a journey to Yellowstone National Park. During your trip, answer the field trip questions and record your answers below. At the conclusion of your journey, you will be asked to make a travel brochure to encourage visitors to come to the park. You may want to take some notes in the space provided.

I. Field Trip Questions

1. What is generally the warmest month(s) in the park? _____

2. During which month(s) does the park get the least rainfall? _____

3. Name the largest geyser in the park. _____

4. What disease has infected many of the bison in the park? _____

5. What is another name for a bison? _____

6. Name the endangered bear species that inhabits the park. _____
   What do these bears eat? _____

7. Moose are the largest deer in the world. They are commonly found in the park. What does a moose eat? _____

8. What type of sheep are commonly seen in the mountainous areas of the park? _____ Why can these sheep live on rock ledges? _____
   _____

9. Wolves have not always lived in the park. They were reintroduced to the park some years ago. This met with great opposition by the ranchers. Why? _____
   _____
   _____
   _____

10. What is the cost of a seven-day pass into the park? _____

11. What state filed a lawsuit against the park that caused it to implement a new bison management plan? _____ What did this new plan include?

_____

_____

II. Notes

III. Developing a Travel Brochure

Fold a sheet of white paper in half. The outside of one side will be your cover. On the cover you should write the name Yellowstone Park and draw an appropriate picture to accompany it. Inside the brochure provide information and pictures that would attract visitors to the park. On the back of the brochure write your name as the travel agent and any other information about the park. Remember to make the brochure informative, colorful, neat, and interesting.

Conclusions Questions

1. What did you enjoy best about the field trip to the park? _____

_____

2. What was the most interesting thing you learned during your visit? _____

_____

3. Now that you have seen the park on the Internet, would you go to the park in person if you had an opportunity? _____ Why or why not? _____

_____

4. Name another place (related to science) that you would like to visit on the Internet.

_____

Outside

TREASURES IN THE FOREST

TIME REQUIRED

One-half to one full school day

Objectives: Students will observe the parts of a forest and learn to appreciate the importance of the forest as a natural resource.

Teaching Strategies: Select an area of the campus where there is a stand of trees. The area you select should contain some wildlife. Avoid locations that contain poisons: ivy, oak, and sumac.

Students will be asked to identify organisms in the forest and the ecotone. Provide some tree identification books. If students do not know or cannot find names of organisms, allow them to describe the organisms instead of naming them.

Caution students to treat the forest and its creatures gently. Picking flowers or walking off the trail damages the homes of many plants and animals.

Divide the class into small groups of three or more. Give each group a clipboard, student activity pages, and a pencil. Individual groups should choose one member to function as the recorder of data.

Instruct students to read the Background Information and answer the Pre-Lab Questions.

Evaluation: At the conclusion of the field trip, each group should turn in the completed student activity pages.

A suggested grading rubric:

Criteria	Points allowed	Points awarded
Pre-Lab Questions correct	20	_____
Students participated in lab work	20	_____
Students remained on task	30	_____
Student activity pages correct	30	_____
Total	100	_____

TREASURES IN THE FOREST

Many campuses have access to a forest or small wooded area. The most common type of forest in our country is the deciduous forest. This type contains trees that lose their leaves in the winter.

Deciduous forests can be divided into four vertical zones: canopy, understory, shrub, and herb layers. The top of the forest, the canopy, is composed of the tallest trees with branches that overlap one another. The understory consists of trees that are shorter than the canopy trees. Shrubs are woody plants with many stems. They typically grow tall enough to reach the lowest branches of the understory trees. The herb layer is made up of short plants, such as the ferns, that grow on the forest floor.

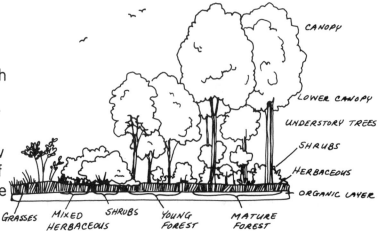

Figure 1. Vertical layers of the forest

The location of a plant in the forest determines how much light it receives. The trees of the canopy receive the most sunlight. If you measure the amount of sunlight from the canopy downward, you find that light intensity decreases. The branches and leaves of canopy trees filter some light, so that by the time it reaches the ground of the forest, it is faint. Plants in the forest live in the places they do because that is where they receive their required amount of light.

The covering on the forest floor, which is called **leaf litter**, is composed of dead leaves and other organic matter. The soil underneath this leaf litter is rich with earthworms and other organisms that help convert organic matter into soil nutrients.

The forest plays different roles to different organisms. It provides meals and homes to a variety of wildlife. Woodpeckers feast on the insects in trees, while redbirds build their nests in the tree branches. Trees furnish these animals with shade, protect them from wind, act as noise and air pollution filters, and supply oxygen. The forest also supplies valuable resources for humans. Pulp made from trees can be used to manufacture paper and wood products. Humans also enjoy many of the fruits produced by trees.

A forest is an **ecosystem,** an area that contains several populations of plants, animals, fungi, and microscopic organisms. These populations are dependent on one another for survival. For example, many forest animals eat plants. Without those plants, they could not survive. When the animals die and decompose, they fertilize the soil. This provides more nutrients for plant growth. It also provides food and habitats for fungi and countless microscopic organisms.

Other ecosystems are streams, lakes, and fields. These differ from one another in the type of organisms they contain. For example, the squirrels that are so common in forests are not found living in fields or in lakes. Similarly, the trees that make up the dominant forest plants are not found in lakes or fields.

Figure 2. An ecotone between a forest ecosystem and a field ecosystem

The place where two ecosystems meet is called an **ecotone**. Ecotones are areas like the edge of the woods or the bank of a creek. In ecotones, you can often find inhabitants of both ecosystems. Turtles that live in lakes sometimes move up on the banks of the lake, and birds that nest in a field may feed along the lake bank.

Pre-Lab Questions

1. Name and describe the layers of a forest. _____

   _____

2. Name three kinds of ecosystems. _____

   _____

3. Describe an ecotone near your school. _____

   _____

4. What does the forest provide for its inhabitants? _____

   _____

TREASURES IN THE FOREST

A. Trees as Resources

Look at the trees that surround you. In the first column, list ten household items that are made from trees. In the second column, indicate what part(s) of the tree the item came from. For example, tree sap is used to make preservatives, and pine cones are used in some Christmas decorations.

Name of Product	Part(s) of Tree Used
1. _____	_____
2. _____	_____
3. _____	_____
4. _____	_____
5. _____	_____
6. _____	_____
7. _____	_____
8. _____	_____
9. _____	_____
10. _____	_____

B. Trees and Other Plants

What are the dominant (most common) trees in this forest? _____

_____

Would you describe this forest as a deciduous forest, evergreen forest, or mixed forest? _____ Why? _____

_____

List or describe the plants that are between two feet and five feet tall.

_____

_____

List or describe the plants that are shorter than two feet.

_____

_____

C. Log Watch
Sit down beside a rotting log, but do not touch it. Observe the log carefully for a few minutes. Then match the following description to the appropriate questions.

Descriptions: a. None present
                  b. Solid and firm
                  c. Few present
                  d. Many present
                  e. Crumbling slightly
                  f. Soft and mushy
                  g. Firm and strong

Questions:

What is the condition of the wood? _____

How many holes are in the wood? _____

How many worm tunnels are in the wood? _____

Are mushrooms or fungi growing on the log? If so, how many? _____

Are mosses growing on the log? If so, how many? _____

Are there plants growing on the log? If so, how many? _____

How many insects can you see on the log? _____

If this log were removed from the forest, would the forest be changed? Explain your answer.

_____

_____

D. Forest Neighbor
Does this forest adjoin another type of ecosystem? If so, what? _____

_____

Describe the ecotone between this forest ecosystem and the adjoining ecosystem. _____

_____

ACTING UP

TIME REQUIRED

One to two hours

Objectives: Students will observe the behavior of earthworms in their natural environment.

Students will evaluate earthworms' behavior in response to a chemical.

Teaching Strategies: Before the trip outside, walk around the campus with a shovel and find out where there are plenty of earthworms. They will probably be living in a moist area under dense leaf litter. On the day of the lab, students will need shovels, water (spray bottles are OK), cotton balls, rulers, and ammonia. Water that has been allowed to stand in an aquarium or bowl overnight will not contain chlorine.

Have students read the Background Information and answer the Pre-Lab Questions.

As an introductory activity, bring one worm to the classroom. Point out the worm's anterior and posterior ends and its clitellum. The anterior end is located closest to the clitellum. Show students the segments on an earthworm and explain that each segment contains a circular muscle. Earthworms also have longitudinal muscles that work with the circular ones and enable them to crawl.

Remind students that they are working with living organisms and they should not hurt them in any way.

Divide the class into small groups of three or more. Give each group a clipboard, student activity pages, and a pencil. Individual groups should choose one group member to serve as the data recorder.

Evaluation: At the conclusion of the field trip, each group should turn in the completed student activity pages.

A suggested grading rubric:

Criteria	Points allowed	Points awarded
Pre-Lab Questions correct	20	_____
Group was on task during the trip	20	_____
Data Table completed	30	_____
Conclusions Questions correct	30	_____
Total	100	_____

ACTING UP

Soil organisms make up a community of living things. Some of the organisms that you might find in soil include millipedes, spiders, and earthworms. All soil organisms depend on their instincts to survive. Instinctively they know how and where to find food, how to reproduce, and how to avoid danger.

Figure 1. A cat responds to danger by arching its back and fluffing out its hair to appear larger to its enemies.

A living thing's response to stimuli in the environment is called **behavior**. A **stimulus** is anything that causes a response from an organism. Behavior can be either inborn or learned. Inborn behavior is involuntary, unlearned, and instinctive. Inborn behavior has predictable responses. For example, if a cat is approached by a scary animal, it arches its back and hisses. Cats are born with this built-in behavior.

Instincts are complex inborn behaviors. Nest building and courtship dances are examples of instincts. Inborn behaviors help animals that do not have highly developed nervous systems to survive and reproduce. These behaviors improve the animals' chances of survival by helping them to respond appropriately to stimuli, even though they have not had the opportunity to learn the best response.

Learned or acquired behavior is gained through experience and involves memory. When you repeat a behavior so often that it is automatic, that behavior is called a **habit**. Habits are different from instinctive behaviors because they are voluntary choices.

The complexity of an animal's nervous system determines its ability to learn behaviors. The more complex the nervous system, the more behaviors the animal can acquire. Simple organisms exhibit more inborn than learned behaviors. Higher animals show more learned behaviors.

Earthworms are simple, segmented organisms that live in loose soil. Their behaviors are instinctive. Small structures on the worm called **setae** provide traction and allow the worm to crawl and burrow into the earth. Earthworms are scavengers that feed on decomposing plant material in the soil.

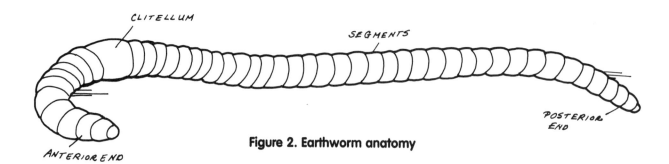

Figure 2. Earthworm anatomy

One way that organisms respond to a stimulus is by moving toward it or away from it. Such movement toward or away from a stimulus is called a **taxis**. Taxis can be positive or negative. Movement toward the stimulus is a positive taxis while movement away is a negative taxis. **Chemotaxis** is response to chemicals, **phototaxis** is response to light, and **geotaxis** is response to gravity.

Pre-Lab Questions

1. Where do earthworms live?_____

2. What is the difference between learned and inborn behavior? _____
   _____

3. Would a simple or a complex animal be more likely to have many inborn behaviors?
   _____

4. Explain why a cat arches its back when it gets scared._____
   _____

5. What does an earthworm eat? _____

6. Define **taxis**. _____
   What is the difference in negative and positive taxis? _____
   _____
   _____

ACTING UP

1. Take a ruler, shovel, water, ammonia, cotton balls, and student activity pages outside.

2. With a shovel, gently move aside some leaf litter and soil until you find two earthworms. Have one lab member gently hold the two worms while the other two prepare the test area.

   - To create a test area, brush the leaf litter away from the soil. With your hands or with the shovel, smooth out a small area of soil about the size of two dinner plates.

   - Divide this area in half by drawing a line through it with a stick.

3. Place the two worms in the test area with their anterior ends resting on the midline.

4. At one end of the test area (in front of the worms) place a cotton ball that has been saturated in ammonia. **Caution. Do not touch the earthworms with any chemical.**

5. Observe the worms for the next 15 minutes.

6. After 15 minutes, measure the distance in centimeters that the anterior end of each worm moved from its original position. If the worm moved forward over the line, record this in the Data Table as a positive number. If it moved backward, record it as a negative number.

7. Repeat the entire procedure (in a different test area), substituting a small pile of soil for the cotton balls soaked in ammonia. Record movement in the Data Table.

Data Table: Movement of worm in response to stimuli.

	Worm 1	**Worm 2**
Test 1: Ammonia in test area		
Test 2: Soil in test area		

Conclusion Questions

1. Describe the location where you found your earthworms: damp or dry? bare soil or leaf litter? sunny or shady? _____

_____

2. In which test did the worms show positive taxis? _____

Negative taxis? _____

3. Why is it important that worms respond as they do to chemicals? _____

_____

4. Did the earthworms demonstrate inborn or learned behaviors? _____ Explain your answer. _____

_____

_____

5. Devise an experiment to demonstrate the response of frogs to heat and cold.

_____

_____

_____

_____

_____

_____

CLASSY CARS

TIME REQUIRED

One to two hours

Objectives: Students will classify 25 cars into eight categories.

Teaching Strategies: Discuss classification with your students prior to this trip. You may want to demonstrate how a classification tree looks and discuss the seven major categories used in the modern biological classification system: kingdom, phylum, class, order, family, genus, and species.

Divide the class into small groups of three. Individual groups should choose one member as the data recorder.

Instruct students to read the Background Information and answer the Pre-Lab Questions.

Take students to the teachers' parking lot and caution them not to touch any vehicles. This activity will be conducted visually. They may not touch any car or open any car doors to look inside.

Evaluation: At the conclusion of the field trip, each group should turn in the completed student activity pages.

A suggested grading rubric:

Criteria	Points allowed	Points awarded
Pre-Lab Questions answered correctly	20	_____
Group was on task during the trip	20	_____
Data Tables I-IV completed	40	_____
Conclusion Questions correct	20	_____
Total	100	_____

CLASSY CARS

Many of our daily activities involve grouping and organizing objects. The technique of grouping items is called **classification**. You probably classify things yourself. Do you keep all of your shorts in one drawer, shirts in a second drawer, and socks in a third drawer? Or do you classify your notes by placing them in separate sections in a notebook?

People commonly use classification at their jobs. The checkout clerk at the grocery store works with different bills and coins that are classified in stacks in the cash register. Sales personnel in clothing stores classify the clothes as Men's, Women's, Children's, and Juniors'. Because the clothing is classified, salespeople know exactly where to place new merchandise when it arrives at the store. Classification is part of our life.

Scientists have been interested in classification for many years. There are millions of different species of organisms known today. Each year this number grows larger as new ones are discovered. A good classification system helps us study living things. Scientists have a special name for the study of classification of organisms: **taxonomy**.

Taxonomists (people who classify living things) have designed five major groups or kingdoms into which they can categorize all living things. Each of these kingdoms has special features that make it unique. As new organisms are found, scientists study their characteristics and place them in the proper groups.

To classify an organism, scientists first place it in one of the five kingdoms: Animal, Plant, Fungus, Protist, or Monera. You are probably most familiar with the Plant and Animal Kingdoms. The Plant Kingdom contains multicellular organisms that can make their own food. The Animal Kingdom is made up of multicellular organisms that can move, but they cannot make their own food. Pine trees fit in the Plant Kingdom while tigers and grasshoppers are members of the Animal Kingdom.

A kingdom is subdivided into smaller groups called **phyla** (phylum in the singular). A tiger belongs to the phylum Chordata; a grasshopper is a member of the phylum Arthropoda.

Phyla can then be divided into classes. Tigers are members of the class Mammalia, warm-blooded animals with mammary glands. Classes can be subdivided into orders, orders can be divided into families, families can be divided into genera (genus in the singular), and genera into species. The genus and the species names of an individual are used together to form an organism's scientific name. This is the most precise division.

To correctly write the scientific name of an organism, one must follow special guidelines. The genus name, which is always written first, is capitalized. The species name is written second, all the letters in lower case. Both names are italicized. For example, the scientific name of the African lion is *Felis leo*. The scientific name of a house cat is *Felis domesticus*. You will notice that both organisms have the same genus because they are both members of the cat family, but they have different species names because they are not exactly alike.

There are three very good reasons for classifying things. First, classifying allows us to group similar things together. This makes it easier for people to study them. Second, classifying lets us characterize individuals. It is important that every kind of living thing have its own, distinguishing name. Third, common names vary from country to country and even from region to region. What one culture thinks of as a "cougar" may be commonly called something else in another culture.

Figure 1.

Felis leo

Felis domesticus

Pre-Lab Questions

1. Name one item that you can find classified in a grocery store. Name one item you can find classified in a department store.

    _____   _____

2. What are the advantages of classifying all living things? _____

    _____

3. What are the largest categories of living things called?_____

4. What are the smallest categories of living things called? _____

5. What is a scientific name? _____

Name _____

CLASSY CARS

I. Record characteristics of 25 automobiles.

In the teachers' parking lot there are a variety of automobiles. You and your team are a special group of taxonomists who have the job of classifying cars into groups.

Select 25 cars. In Data Table I, record the color, make, tag number, and special features of these cars. Record all of your data about the cars in Data Table I. One special rule is that you cannot touch any of the automobiles.

Data Table I

Car # (include parking space number if available)	Color of car	Make or type of car	Tag number	Special features you notice about the car
#1				
#2				
#3				
#4				
#5				
#6				
#7				
#8				
#9				
#10				
#11				
#12				

#13				
#14				
#15				
#16				
#17				
#18				
#19				
#20				
#21				
#22				
#23				
#24				
#25				

II. Divide into two large groups.

Use the characteristics that you have listed to divide the 25 cars into two groups. Record these two groups in Data Table II. Select a name for each group. Below the table, write the criteria you used to separate the cars into these two groups.

For example, you might classify the cars based on the number of doors they have. You could name the first group "Two-Door Cars" and the second group "Four-Door Cars." Under the heading of "Two-Door Cars," you would list the numbers of all cars that fit into that group. Do the same for the four-door cars. For criteria you would write: Number of doors.

Data Table II. Two major groups of cars in the parking lot

Group Name: _____	Group Name: _____

Criteria you used to distinguish the cars into these two groups: _____

III. Subdivide the cars into four categories.

Take the cars in Group 1 and separate them into two groups. Do the same for the cars in Group 2. When you finish, you will have four categories. Name each category and record the number of cars in each category in Data Table III. At the bottom of the table, cite the criteria you used to separate these cars.

For example, you might divide the Group 1 (the two-door cars) cars into light-colored cars and dark-colored cars. You could do the same for Group 2. You could name these divisions: Two-Door Light, Two-Door Dark, Four-Door Light, and Four-Door Dark.

Data Table III. Four categories of cars

Group: _____	Group: _____	Group: _____	Group: _____

Distinguishing criteria used to group these cars: _____

IV. Form eight groups.

Take each of your existing four groups and separate them into two categories. This will give you eight groups of cars. Record this information in Data Table IV. Name each group and describe what criteria you used to separate them. You might wish to separate the cars based on such characteristics as those that have bumper stickers and those that do not.

Data Table IV. Eight categories of cars

Group: _____	Group: _____	Group: _____	Group: _____	Group: _____	Group: _____	Group: _____	Group: _____

Distinguishing factor used to make the eight groups: _____

Conclusion Questions

1. Name one business where cars are classified. What basis is used to classify cars at that business? _____

2. If you were asked to classify the cars in this activity one step further, how many groups would you have? _____

3. Explain why you were called **taxonomists** in this activity. _____

_____

_____

4. List five locations at the school where things are classified. Name what is classified at those locations and what criteria is used to separate the items into groups.

_____

_____

_____

_____

5. Music is often classified in stores. If you worked as a clerk in a music store, how would you classify the CDs?

_____

_____

_____

SEED HUNT

TIME REQUIRED

Day 1 (Parts I and II)—One to two hours

Days 2–60 (Part III, optional)—5 minutes each day

Objectives: Students will collect seeds from the school yard and determine how they are dispersed.

Students will plant the seeds they collect in a small garden and follow their growth over the next several weeks.

Teaching Strategies: Discuss the different ways that seeds can be dispersed over an area. A day or two before the lab, ask students to bring in old wool socks, flannel material, or any type of fuzzy cloth or old clothing. These will be pulled across an open field so that seeds will cling to them.

Divide the class into groups of two or three. Individual groups should choose one group member to serve as the recorder of data.

Instruct students to read the Background Information and answer the Pre-Lab Questions.

On the day of the lab, take students outdoors and dictate the boundaries of the area in which they should work. Caution students not to leave the designated area. You may want to point out any poisonous plants in this area prior to the activity.

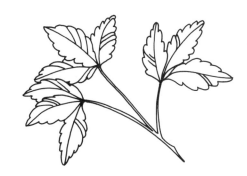

Figure 1. Poison ivy

Evaluation: At the conclusion of the field trip, each group should turn in the completed activity pages.

A suggested grading rubric:

Criteria	Points allowed	Points awarded
Pre-Lab Questions correct	20	_____
Group was on task during the trip	20	_____
Data Table I completed	20	_____
Data Table II completed	20	_____
Questions answered correctly	20	_____
Total	100	_____

SEED HUNT

A seed contains an embryo plant and the food that plant needs for its early development. Some plants produce structures that contain many seeds. For example, a pine tree has female cones that bear many seeds. These seeds are protected by the scales of the cone. If you carefully examine the seed of a pine cone, you find that it has a "wing" attached to it. The function of the wing is to help the seed travel on gusts of wind.

Most plants reproduce by forming seeds. When these seeds find a suitable location, they germinate. **Germination** is the development and growth of a baby plant. If all of the seeds produced by one plant fell to the ground and germinated near the parent, the young plants would be crowded. Overcrowding prevents all of the germinating plants from surviving. Consequently, plants have developed methods of spreading the seeds away from the parent. The traveling or scattering of seeds to various locations is called **dispersal**.

Some, but not all, seeds are contained in fruits. A fruit is a plant's mature ovary. Seeds within fruits are called **covered seeds**. A blueberry is an example of a fruit that contains seeds. Seeds that are not contained in fruits are referred to as **naked seeds**. The pine tree is an example of a plant that produces naked seeds on a cone instead of in a fruit. Both covered and naked seeds can be dispersed by water, wind, or animals.

Figure 1. A female pine cone and a seed from that cone

Lightweight fruit and seeds can be easily dispersed by the wind. Dandelions are an example of such a fruit. The fruit of the dandelion is made of a collection of fibers that form a "parachute-like" design. A small puff of wind can carry these seeds for miles. Maple trees produce fruits that are shaped like tiny propellers that can float through the air.

Figure 2. A dandelion seed is contained in a lightweight fruit that is shaped like a parachute.

A coconut is a large, heavy fruit. If it drops from the tree into water, it can float to a new location. Plants that live on hillsides often have their seeds dispersed as rain washes them to the valley below.

Seeds can also be dispersed by animals that unknowingly carry them internally or externally. Since certain fruits, such as apples and blackberries, are tasty to animals, they are consumed as food. Many times an animal carries a fruit long distances before consuming it and drops seeds on the ground in the process. Other animals may eat the fruit and its seeds. Because seeds are very difficult to digest, they often stay intact and pass all the way through the digestive tract of the animal. When the animal deposits its feces, the seeds are released and eventually germinate.

Other seeds travel on the outside of animals. These have small barbs or hooks on them that attach to the animal's fur like a hitchhiker. Later these seeds may drop off the animal miles from their original location. If they fall in a favorable location, seeds can germinate.

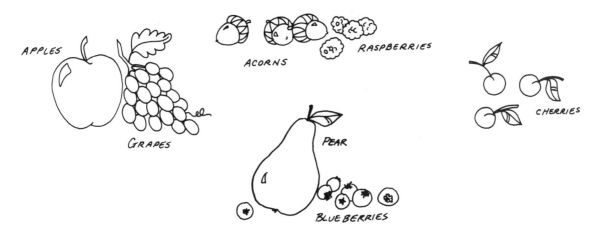

Figure 3. Some fruits commonly eaten by animals

Pre-Lab Questions

1. What is a seed? _____

_____

2. What is the difference in a naked seed and a covered seed? _____

_____

3. What is a fruit? _____

4. List three ways that seeds can be dispersed._____

_____

_____

5. Give examples of two plants whose seeds are dispersed by wind._____

_____

Name _____

SEED HUNT

Your teacher will give your group a piece of cloth or an old wool sock to help you collect seeds.

I. Collecting seeds

Collect plant seeds from the locations your teacher specifies. Some seeds can merely be picked from plants or gathered from the ground. Other seeds may be obtained by dragging a cloth over an area of plants. Use both techniques to gather your seeds.

After you find a seed, draw it, describe it, explain where you found it, and describe how you think it is dispersed in nature. Place this information in Data Table I.

Data Table I

Drawing of the seed	Description of the seed	Location from which obtained	Probable mechanism for dispersal

II. Creating a new method of seed dispersal

Use your imagination to invent a new plant. Draw and describe your plant in the space below. Also sketch the seed and/or fruit that plant produces. Invent a new method of seed dispersal for this plant (do not use the wind, water, or animals). When designing your plant and its seed or fruit, be certain that the structure of the seed matches your dispersal choice. Write a description of how your seed will be dispersed.

Data Table II

Drawing of your new plant	Drawing of its seed/fruit	Description of seed/fruit dispersal

III. Planting seeds—optional

In the location specified by your teacher, plant the seeds that you gathered. On small wooden stakes (such as Popsicle sticks) use permanent markers to draw a picture of each kind of seed. Place the stakes in the ground near the appropriate seeds.

For the next month or two follow the progress of your seeds and see what kinds of plants develop.

DIGGING IN THE DIRT

TIME REQUIRED

Two to three hours

Objectives: Students will compare the quality of soil in different locations on campus.

Teaching Strategies: Prior to the trip have students read the Background Information and answer the Pre-Lab Questions so they will have a general understanding of terminology associated with soil study.

Divide the class into groups of three or more students. Give each group the following: small shovel, large container of water, hand lens, ruler, soil pH kit, clipboard, student activity pages, and a pencil. Individual groups should choose one group member to serve as the data recorder.

Instruct groups to select four different locations on campus for soil study. Recommend that students choose some sites on campus that are well traveled and some that are not. Additionally, some of the areas they select should have plant life, and others should not. Suggestions for experimental sites include planting beds, playground, forest, hillside, open field, and paths.

Before going outdoors, show the class how to use the soil pH kit to test the soil. Each kit contains directions.

Evaluation: At the conclusion of the field trip, each group should turn in the completed student activity pages.

A suggested grading rubric:

Criteria	Points allowed	Points awarded
Group was on task during the trip	20	_____
Student activity pages (Sections I and II) completed	50	_____
Creative story (Section III) completed	30	_____
Total	100	_____

DIGGING IN THE DIRT

Soil is made up of several ingredients: bits of sand, silt, clay, and humus. The sand, silt, and clay particles are the result of rock weathering. Rocks are broken into smaller pieces by the physical and chemical forces of weather. The humus component of soil comes from dead plants and animals. A mature, fertile soil is the result of hundreds of thousands of years of accumulating decayed plants and animals and rock weathering.

Not all soils are the same. Generally, soils in cold, dry climates are shallow and do not contain as much humus as soils in wet, warm climates. That is because the chemical reactions that produce humus and weather rocks occur faster in warm areas. Also, plants grow faster in warm weather than they do in cool. Therefore, there is more plant mass available for decay.

The top layer of soil is often referred to as topsoil. Usually, it is more fertile than the layers under it. Topsoil's fertility is due to the humus, which is rich in plant nutrients. Beneath the topsoil is the subsoil, which contains more minerals and less humus.

Soil that is loosely arranged and that contains a variety of minerals and nutrients often supports animal life. Animals that live within the soil get their food, water, and air from the soil. Plants have similar needs; their roots must be able to penetrate the soil to find food, water, and air. That is why there is less animal and plant life found in tightly packed soil than in loosely packed soil. Every time we walk or drive across soil, we pack it a little more.

Pre-Lab Questions

1. Which soils contain the most humus: soils in warm climates or those in cool climates? _____ Why? _____

   _____

2. What is topsoil? _____
   Why is it more fertile than subsoil? _____

   _____

3. Do you think you would be more likely to find soil animals in tightly compacted soil or in soil whose particles are loosely arranged? _____

   Why?_____

   _____

   _____

DIGGING IN THE DIRT

Collect the materials your teacher has provided for you to take outside. At each site, complete Parts I and II below.

I. Site Descriptions. Write a descriptive phrase or sentence about each of your sites. Examples: "On playground, near swings" or "Near the oak tree on front campus"

Site 1: _____

Site 2:_____

Site 3:_____

Site 4:_____

II. Complete the Data Table for each site. Finish all of the activities at Site 1 before moving to a new location. The letters on the chart match the activity letters.

Activity A—Dig a hole about four inches in diameter and three inches deep in the soil. Remove the soil and observe its color and texture. Record this information on the Data Table.

Activity B—Inspect the soil at this site for animal life: millipedes, ants, earthworms, etc. Use the hand lens, if necessary. List what you see. If you do not know the names of animals you see, describe them.

Activity C—Are there any plants or plant parts at this site (decayed leaves, twigs, seeds, etc.)? If so, list the parts you see.

Activity D—Pick up a handful of dirt and squeeze it as tightly as possible. When you open your hand, does the soil crumble apart or stick together? Would you describe the soil as dry or wet?

Activity E—Pour enough water in the hole to fill it. After five minutes, use your ruler to determine the number of centimeters the water level has dropped from the top of the hole.

Activity F—Use the pH kit to determine the soil pH.

Replace the clump of soil you removed from the ground.

Data Table

	Site 1	Site 2	Site 3	Site 4
A. Color and texture of soil				
B. Animal life present				
C. Plant life present				
D. Result of the squeeze test				
E. Result of the water penetration test				
F. Result of the pH test				

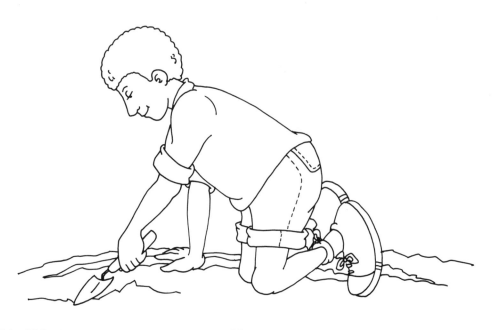

III. Pretend that you are a little plant that needs plenty of water and nutrients. In the space below, write a short story explaining in which of the four sites you would prefer to live. Express your views from the standpoint of the plant and discuss what you need to grow.

Conclusion Questions

1. Of the four sites you tested, in which was the soil the most compacted? _____
   Why do you think this is so?_____

   _____

2. Of the four sites you tested, which contained the most moisture? _____
   Why do you think this is so?_____

   _____

3. Based on your results today, do you think you would be more likely to find a lot of soil animals living in soil from a path, or soil in an area where people do not walk?
   _____ Why?_____

   _____

4. Based on your results today, through what type of soil does water drain faster: soil on a path or soil in the woods? _____ Why? _____

   _____

   _____

Off Campus

SEWAGE SAFARI

TIME REQUIRED

One-half to one full school day

Objectives: Students will visit a sewage treatment plant and observe the steps in sewage purification.

Teaching Strategies: Prior to the trip to the sewage plant, discuss the importance of water and sewage treatment. Explain to students the relationships of municipal sewage and water treatment plants. Include a discussion of bacterial diseases that can spread through untreated water. Remind students that stream water is not treated, so campers should always sterilize stream water or consume bottled water.

Divide the class into small groups of three or more. Each group should be supervised by a teacher or chaperone. Give each student a clipboard, activity page, and a pencil.

Instruct students to read the Background Information and answer the Pre-Lab Questions so they will have some general knowledge of the terms discussed on the student activity pages. If you find any words especially difficult, you may want to discuss these prior to the field trip.

Instruct students to complete the student activity pages as they tour the facility. If they have specific questions, there will be an opportunity for them to ask questions of the plant manager.

Evaluation: At the conclusion of the field trip, each group should turn in the completed student activity pages.

A suggested grading rubric:

Criteria	Points allowed	Points awarded
Pre-Lab Questions correct	20	_____
Group was on task during the trip	20	_____
Student activity page questions answered	40	_____
Well-done follow-up activity	20	_____
Total	100	_____

SEWAGE SAFARI

All communities have to deal with disposal of liquid waste material, called **sewage**. Sewage includes water from sinks, toilets, and washing machines. To prevent it from contaminating the drinking water supply, sewage must be handled properly. Each community has a treatment plant that is designed to cope with sewage.

Not all sewage treatment facilities are the same: some are more effective than others. There are three basic methods of sewage treatment: mechanical, biological, and chemical. Sewage treatment facilities use some or all of these methods.

Mechanical treatment is also referred to as **stage 1** or **primary treatment**. During this phase, large particles are separated from the liquid by a filtering device (Figure 1). The filtered liquid is pumped into a large tank. In this tank, heavier molecules fall to the bottom where they collectively form a thick deposit called **sludge**. This sludge is later removed. If any oil is present in the sewage, it is skimmed from the top during this stage.

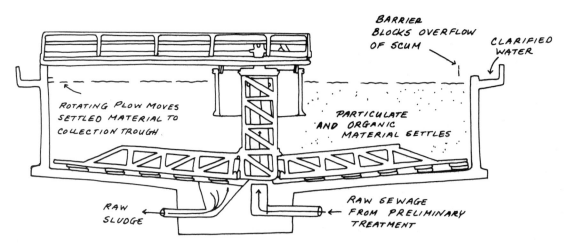

Figure 1. Primary treatment

Secondary or **biological treatment** occurs in the same large tanks. During this process living organisms act on the sludge to break it down. These organisms are usually sludge-eating bacteria. As the sludge is broken down, methane gas is produced. It exits through special pipes where it can be collected and used for fuel. After treatment with bacteria, the sludge is filtered from the tank. It can be dried and used for fertilizer or placed in a landfill.

The final process, called **tertiary treatment** (Figure 2), is accomplished with chemicals. Chlorine or ozone is added to the treated sewage to disinfect it. Then it is tested to see whether it is clean enough to release. Treated sewage can be sprayed onto trees or crops, or released into streams.

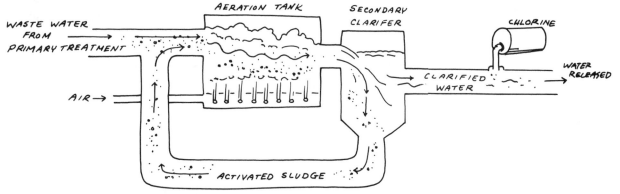

Figure 2. Tertiary treatment

Pre-Lab Questions

1. Where does sewage come from? _____

    _____

2. Why must sewage be treated before it can be released into creeks or sprayed on crops?

    _____

    _____

3. Name and describe the three stages of sewage treatment._____

    _____

    _____

    _____

    _____

4. What is sludge? _____ What happens to the

    sludge produced in a sewage treatment plant? _____

    _____

    _____

5. What happens to the methane gas produced by the breakdown of sludge?

    _____

    _____

SEWAGE SAFARI

As you tour the sewage treatment facility, complete this student activity page. Some questions may have to be answered by the plant manager. There will be an opportunity during your visit for you to ask questions.

Name of sewage treatment facility:

Name of tour guide and his/her job description:

1. Name the steps or stages of treatment used by this facility to purify wastewater.

   _____

2. What is the first step that sewage undergoes in treatment at this facility?

   _____

3. Where does this sewage plant send its sludge? _____

   _____

4. Name some tests that this facility conducts on the wastewater before it is released.

   _____

5. If this plant includes the biological treatment phase, name the organism it uses in this process. _____ Does this organism require oxygen to function?

   _____

   _____

6. What does the sewage treatment plant do with gases that are produced by the breakdown of sludge?

   _____

7. List five interesting facts you learned on your trip today.

   a. _____

   b. _____

   c. _____

   d. _____

   e. _____

II. Follow-up Activity:

Assume you are the plant manager of the sewage treatment facility. Recent news reports have stated that many communities are doing a poor job purifying the water. Write a one- or two-paragraph news release that will be televised on the evening news. Explain to the concerned citizens of your community that their sewage is properly treated.

AQUARIUM ADVENTURES

TIME REQUIRED

One-half to one full day.

Objectives: Students will visit a public aquarium and record information about animals found there.

Teaching Strategies: Before the trip, discuss the difference in vertebrates and invertebrates. Describe the classes of vertebrates—mammals, reptiles, birds, amphibians, and fish. Draw a fish on the board and label its parts. Show the class where the gills are located and explain how gills remove oxygen from the water.

Divide the class into small groups of three or more. Each group should be supervised by a teacher or chaperone. Give each group a clipboard, student activity pages, and pencil. Have one member of the group record data.

Have students read the Background Information and answer the Pre-Lab Questions. Instruct them to complete the student activity page during the field trip.

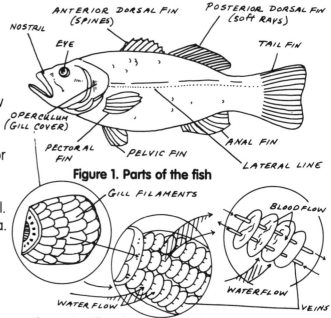

ANTERIOR DORSAL FIN (SPINES)
POSTERIOR DORSAL FIN (SOFT RAYS)
NOSTRIL
EYE
TAIL FIN
OPERCULUM (GILL COVER)
PECTORAL FIN
PELVIC FIN
ANAL FIN
LATERAL LINE

Figure 1. Parts of the fish

GILL FILAMENTS
BLOOD FLOW
WATER FLOW
WATER FLOW
VEINS

Figure 2. Gills take in oxygen from water.

Evaluation: At the end of the field trip, each group should turn in completed student activity pages.

A suggested grading rubric:

Criteria	Points allowed	Points awarded
Pre-Lab Questions correct	20	_____
Group was on task during trip	20	_____
Completion of student activity pages	20	_____
Aquarium alphabet student page correct	20	_____
Story on favorite animal	20	_____
Total	100	_____

AQUARIUM ADVENTURES

Animals that spend all or part of their time in the water can be found at an aquarium. Inside the aquarium you will see large tanks of water that contain fish and other aquatic animals. The water tanks are especially designed to match the natural living conditions of the animals. The temperature and chemical balance of the water in the tank is carefully controlled.

When people study fish, they usually classify them into groups. There are two major types of fish that you may see at the aquarium: marine and fresh water. Fish that live in salt water are described as marine, and those in lakes and streams are called **freshwater fish**.

Another way that fish can be grouped is by their skeletal type. Some fish have skeletons made of cartilage, the same kind of tissue that gives shape to your outer ear. Cartilage is not as hard as bone. Examples of fish with cartilaginous skeletons are sharks, rays, and skates. Rays and skates have flat bodies with fins that look like wings. In the tropical seas, skates and rays spend most of their time buried in the sand. For protection, rays have a poisonous spine on the tips of their tails.

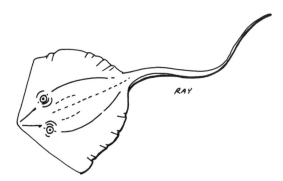

Figure 3. A ray has a skeleton of cartilage and a poisonous spine on its tail.

Most other fish have skeletons of bone. Bony fish are considered to be more advanced than cartilaginous ones. They can be found in all of the fresh and marine waters of the world.

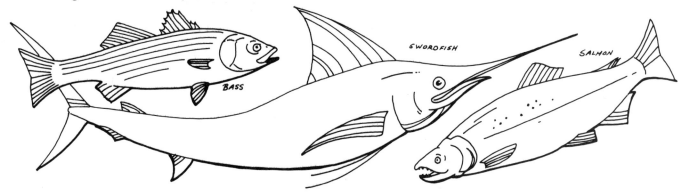

Figure 4. Swordfish, bass, and salmon are bony fish.

Fish are perfectly adapted for life in the water. They can swim with little resistance because they have streamlined bodies. This means that they have a narrow head that enlarges to a wider midsection and then tapers to a narrow tail. This shape allows the water to flow smoothly over the fish's body. As the fish moves, water enters its mouth and flows over its gills, which are lined with small blood vessels called **capillaries**. Oxygen diffuses from the water into the capillaries. At the same time, carbon dioxide leaves the capillaries and passes into the water. Water leaves the fish's body through the gill slits located beneath the flap of the gill.

On your tour of the aquarium, you will find that some tanks contain both vertebrates and invertebrates. **Invertebrates**, like jellyfish, starfish, sponges, and coral, do not have backbones. Generally, invertebrates lack the higher levels of organization seen in vertebrates. However, these animals have specialized traits that help them survive.

Figure 5. Some of the invertebrates that live in water are starfish, sponges, and jellyfish.

Vertebrates are animals with backbones, such as fish and crocodiles. Vertebrates can be divided into five major groups: mammals, birds, reptiles, amphibians, and fish. Besides fish, some of the other vertebrates spend part of their time in water. Amphibians, such as frogs and salamanders, lay their eggs in water. However, many of the adult forms live on land. Reptiles, such as alligators and turtles, live in the water, even though they do not have gills and have to swim to the surface to breathe. The water helps keep them cool and provides them with food. Many species of birds have long legs for wading and spear-shaped beaks for catching fish. Some marine mammals include whales and dolphins. These animals are so highly adapted to the water that they cannot leave it.

Pre-Lab Questions

1. What is the difference between marine fish and freshwater fish? _____
   _____

2. Give examples of three marine invertebrates and three marine vertebrates. _____
   _____
   _____
   _____

3. What are the five major groups of vertebrates? Give a water-dwelling example from each group.

   _____   _____

   _____   _____

   _____   _____

   _____   _____

   _____   _____

4. What do we mean when we describe a fish's body as "streamlined"? _____
   _____

   What is the advantage of a streamlined body to fish? _____
   _____

AQUARIUM ADVENTURES

I. Completion. As you tour the aquarium, complete the following questions and statements about animals that you see on your visit today. Be specific when you record your answers by giving the full names of the animals.

1. a fish that lives in fresh water _____

2. a fish that lives in salt water _____

3. the names of three vertebrates _____,
_____, and _____.

4. the names of three invertebrates _____,
_____, and _____.

5. a mammal _____.

6. a reptile _____.

7. an amphibian _____.

8. What was your favorite aquarium animal? _____ Write two to
three sentences that describe it. _____

_____

_____

_____

II. Aquarium Alphabet. Write down the name of an animal that you see today at the aquarium for each letter of the alphabet. Write something you observed about that animal beside its name.

	Name of animal and a sentence about it
A	Example: Alligator lies in shallow water with nostrils above water; can move quickly when it wants to.
B	
C	
D	
E	
F	
G	
H	
I	
J	
K	
L	
M	

N	
O	
P	
Q	
R	
S	
T	
U	
V	
W	
X	
Y	
Z	

III. **Your Group's Story.** After you have completed your tour of the aquarium, look back at the animal which you and your group selected as your favorite. As a group, write a two-paragraph story in which you become that animal. Describe all facets of your life, such as your appearance, your home, what you eat daily, and your habits.

_____
_____
_____
_____
_____
_____
_____
_____
_____
_____
_____
_____
_____
_____
_____
_____
_____
_____
_____
_____
_____

IT'S A ZOO OUT THERE!

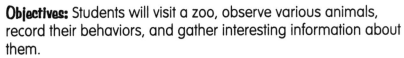

Teacher Information

TIME REQUIRED

One-half to one full school day

Objectives: Students will visit a zoo, observe various animals, record their behaviors, and gather interesting information about them.

Teaching Strategies: The day before lab, place an asterisk (*) by the name of a different animal on each group's activity page. This will be the animal assigned to that group to research during the field trip.

Divide the class into small groups of three or more. Each group should be supervised by a teacher or chaperone. Give each group a clipboard, student activity pages, and a pencil. Individual groups should choose one member to function as the data recorder.

Instruct students to read the Background Information and answer the Pre-Lab Questions so they will have some general knowledge of the terms discussed on the student activity pages.

Have each group tour the zoo with their supervisor and complete the activity page as they view the various animals. Remind them to pay special attention to the animal labeled with an asterisk. Point out the section for the assigned animal at the end of the activity page. Remind students that they will be asked to pantomime the behavior of their assigned animal in front of the class. The observing group that correctly guesses the name of the animal being pantomimed wins a free token. The group with the most tokens at the end of the session receives a prize. (This prize can be free homework passes, candy, etc.) After pantomiming, the group data recorder should read all of the information collected about this animal. Remind groups to keep the names of their assigned animals secret from other groups.

Included on the student activity pages is a list of animals that students may view at the zoo. You can modify this list to fit your situation.

Evaluation: At the end of the field trip, each group should turn in the completed student activity pages.

A suggested grading rubric:

Criteria	Points allowed	Points awarded
Pre-Lab Questions completed	20	_____
Group was on task during the trip	30	_____
Student activity pages completed	30	_____
Pantomime and description appropriate	20	_____
Total	100	_____

IT'S A ZOO OUT THERE!

Watching the animals at the zoo is fun. While watching, you can learn some interesting information about them. Below are some terms you should know:

Adaptations

Structures or behaviors that animals exhibit which help them survive. For example, the webbed feet of a duck are an adaptation that helps it swim (see Figure 1). The lizard's ability to change colors is an adaptation that protects it from its enemies.

Herbivore, Carnivore, and Omnivore

Terms that describe the diets of animals. Herbivores eat plants. Carnivores eat meat. Omnivores eat both plants and meat. Deer are herbivores, lions are carnivores, and bears are omnivores.

Habitat

The place where an animal lives. The habitat of a millipede is a rotting log (see Figure 2). The habitat of an owl is a forest.

Fish

Swimming vertebrates (animals with backbones) that use gills to remove oxygen from the water. Examples of fish are tuna and bass.

Amphibians

Vertebrates that spend part of their life on land and part of it in water. They have moist skin and lay their soft eggs in water. Frogs and salamanders are amphibians (see Figure 3).

Figure 1. A duck's webbed feet are a special adaptation for swimming.

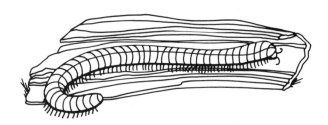

Figure 2. The habitat of a millipede is a rotting log.

Figure 3. An amphibian, such as a frog, lives part of its life in water and part of it on land.

Reptiles

Land vertebrates that have dry skin and lay eggs covered with a tough, leathery shell. Lizards and turtles are reptiles.

Birds

Land vertebrates that are covered with feathers. They lay eggs that are protected with a hard shell. Robins and hawks are birds.

Mammals

Land vertebrates that are covered with hair. They feed their young with milk from their mammary glands. Humans and cats are mammals.

Pre-Lab Questions

1. Write each animal's name in the correct part of the Data Table.

   Whale—lives in water; nurses young with milk from mammary glands
   Gila Monster—lays eggs on land; body covered with dry, hard skin
   Ostrich—lives on land; lays eggs covered with brittle shell
   Toad—spends part of life in water; lays soft eggs in water
   Newt—can absorb oxygen from water or from air; lives in water most of life
   Platypus—swims in water; builds nests on land; feeds milk to young
   Flying Squirrel—covered with hair; nurses young from mammary glands
   Brim—lives in water, its gills absorb oxygen from water

Mammals	Birds	Reptiles	Amphibians	Fish

2. Describe your habitat. _____

   _____

3. What are some special adaptations of polar bears?_____

   _____ fish? _____

   _____

IT'S A ZOO OUT THERE!

STUDENT ACTIVITY PAGE

As you walk around the zoo, complete the following data table. Your teacher may wish to assign you a specific animal. Be certain to observe it closely and complete the detailed information at the bottom of the next page. Your group will be asked to pantomime this animal's behaviors for the other groups.

Data Table

Animal	Habitat	Omnivore, Herbivore, Carnivore?	Mammal, Bird, Fish, Reptile, Amphibian?	Special Adaptations	Description of behaviors you observed
Example—Bat	Caves, Forests	Omnivore	Mammal	Radar-like sense, wings	Hangs upside down
Rhinoceros					
Lemur					
Flamingo					
Zebra					
Sea Lion					
Elephant					
Lion					
Eagle					
Pot-Bellied Pig					
Tiger					
Donkey					
Black Mamba					
Tree Frog					

Sheep					
Llama					
Orangutan					
Alligator					
Boa Constrictor					

Information about your group's assigned (*) animal:

Name of animal _____

Detailed description of its behavior (for 15 minutes):
    Minute 1:
    Minute 2:
    Minute 3:
    Minute 4:
    Minute 5:
    Minute 6:
    Minute 7:
    Minute 8:
    Minute 9:
    Minute 10:
    Minute 11:
    Minute 12:
    Minute 13:
    Minute 14:
    Minute 15:

This animal is a native of _____

Specific foods eaten by this animal: _____

Additional information of this animal:_____

A MOOOOING EXPERIENCE

TIME REQUIRED

One-half to one full school day

Objectives: Students will visit a dairy farm and determine how milk gets from the cow to the store.

Teaching Strategies: Prior to the trip to the dairy farm, make certain students read the Background Information and answer the Pre-Lab Questions.

Divide the class into small groups of three or more. Each group should be supervised by a teacher or chaperone. Give each group a clipboard, activity page, and a pencil. Individual groups should choose one group member to serve as the recorder of data.

Instruct groups to tour the dairy farm with their supervisor and complete the student activity pages as they view the animals.

Evaluation: At the conclusion of the field trip, each group should turn in their completed student activity page.

A suggested grading rubric:

Criteria	Points allowed	Points awarded
Pre-Lab Questions correct	20	_____
Group was on task during the trip	20	_____
Student activity page correct	30	_____
Calculations performed correctly	30	_____
Total	100	_____

A Mooooving Experience

Milking techniques have changed a lot in the last 500 years. When the first cows were brought to our country, milking was performed by hand. The farmers sat on stools and collected milk in a bucket by squeezing the cow's teats. Teats are extensions of udders, the organs in which cows make and store milk.

The udder is divided into four separate quarters. Each quarter has its own milk supply. When the udder becomes full of milk, each section can be emptied through one teat. Today many large dairy farms speed up this process by using automatic milking machines and computers. These machines improve milk production and help to ensure that the milk is free of harmful bacteria.

Cows produce milk to feed their calves. When the cow is pregnant, the udder begins to take raw materials from the bloodstream to make milk. Later a special hormone causes the udder to secrete milk as the calf is almost ready to be born. As long as the calf nurses or the cow is milked, the udder will produce milk. Over a period of time the milk production will decrease, and the cow will have to be rebred to begin the process again.

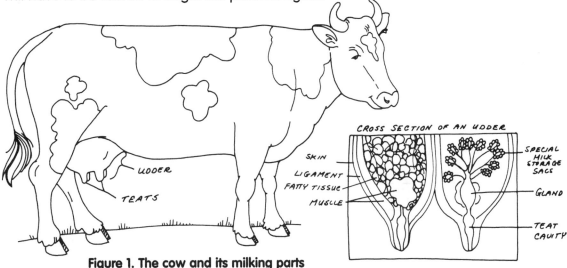

Figure 1. The cow and its milking parts

As the cow is about to be milked or nursed, another hormone is discharged into the bloodstream. This hormone is triggered by stimuli such as the washing of the udder prior to milking or the nuzzling of the cow's teat by the calf. The release of this hormone causes structures in the udder to deliver the milk into ducts. Milk in the ducts stimulates the muscle between the udder and teat to relax. As the machine or farmer gently squeezes the teat, milk passes out of the cow.

Sometimes a cow will not yield milk during the milking process even though she is producing milk. This can happen because the cow gets upset or frightened. Hormones produced as the result of fear keep the milk from passing from the udder into the teats. Cows must be handled gently prior to the milking process.

Pre-Lab Questions

1. Before the milking machine was invented, how did farmers milk cows?_____
   _____

2. Describe the structure and function of an udder. _____
   _____
   _____

3. Why do you think that milk collected by milking machines contains fewer bacteria than milk collected by hand? _____
   _____
   _____

4. Why do cows produce milk? _____
   _____
   _____

5. What chemical signal causes the release of milk into ducts, then into teats? How is this signal triggered? _____
   _____
   _____

6. Why must dairy cows be treated gently? _____
   _____
   _____

A MOOOOVING EXPERIENCE

I. Completion. As you tour the dairy farm, complete the following sentences and questions. Be specific when you record your answers.

1. What color(s) are the dairy cows on the farm? _____ What is the name of the breed of cows? _____

2. Why is this particular breed of cow used on dairy farms? _____

   _____

3. How many pounds (approximately) of food does a cow eat each day? _____

4. Why does the farmer wash the cow's udder with warm water prior to milking?

   _____

5. When milking machines are attached to the cows' teats, milk passes through lines into a(n) _____ _____. Discuss the temperature of this device: _____

6. Milk is transported to a processing plant where it is pasteurized. Pasteurization is a method of killing _____ _____ in milk by rapidly heating and cooling it.

7. After milking, about how long does it take the distributor to get the milk pasteurized, packaged, and delivered to the grocery store shelf? _____

8. What is one disease that may be common to dairy cows? _____

   What are some symptoms of the disease? _____

9. Which states are the leading milk producers? _____

10. What animals, other than cows, are used to produce milk for human consumption?

    _____

11. How many workers does it take to operate the dairy farm? _____

A CAVE RAVE

Teacher Information

TIME REQUIRED

One-half to one full school day

Objective: Students will visit a cave and identify formations within the cave.

Teaching Strategies: As an introductory activity about a week before the field trip, have students grow some crystals in class. These crystals will resemble structures they will see inside caves. Instructions below can be copied and given to each lab group.

> In a beaker, heat about two cups of water on a hot plate.
> Add a ½ cup of Epsom salts to the warm water.
> Stir to dissolve. Place two charcoal briquettes in the solution.
> Add a few drops of red food coloring to the briquettes.
> Place the beaker of this solution in a safe place.
> Observe daily.

In about five days, students will find that crystals of Epsom salts (magnesium sulfate) have grown on the briquettes. Discuss how these are similar to structures seen in caves.

Prior to the trip, have students read the Background Information and answer the Pre-Lab Questions.

Divide the class into small groups of three or more. Each group should be supervised by a teacher or chaperone. Give each group a clipboard, student activity pages, and a pencil. Individual groups should choose one group member to function as the recorder of data. Instruct groups to tour the cave with their supervisor and complete the student activity pages as they view the various structures.

Evaluation: At the conclusion of the field trip, each group should turn in the completed student activity pages.

A suggested grading rubric:

Criteria	Points allowed	Points awarded
Pre-Lab Questions correct	20	_____
Group was on task during the trip	20	_____
Activity pages completed correctly	30	_____
Crystals grown in group	30	_____
Total	100	_____

A CAVE RAVE

Mountains are natural land formations that are built slowly over a long period of time. There are several different mechanisms that contribute to mountain-building. One way that mountains were made was by the accumulation of skeletons of small marine creatures on the ancient sea floor. These skeletons were pressed together into a type of rock called **limestone**. The chemical composition of limestone is calcium carbonate.

As time passed, rocks were later deposited in layers on top of the limestone. The seas receded and these rock layers hardened. Earthquakes pushed the layers of rock upwards. As these rock layers rose from the ocean floor, they cracked.

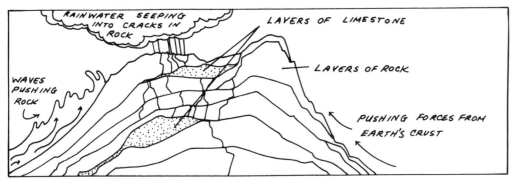

Figure 1. Formation of underground limestone deposits in mountains

Rain water reached the inner layers of limestone by seeping through the cracks. Rain water is naturally acidic because it contains carbonic acid. Carbon dioxide gas and water vapor in the air combine to form carbonic acid. This acid water reacted slowly with the limestone in the rocks, dissolving it. As time passed more and more limestone was dissolved, resulting in huge underground cavities that we call **caves**.

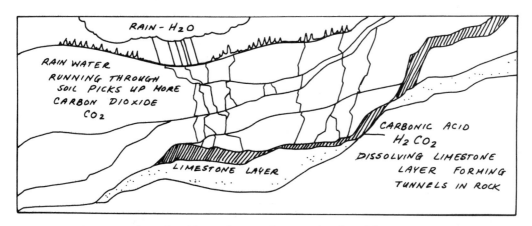

Figure 2. Carbon dioxide and water form carbonic acid.

Inside caves you can usually see structures called **cave deposits** on the walls, floor, and ceiling. The process is very slow, and formation of a deposit only a cubic inch in size can require more than 100 years. The formation of these deposits is affected by the amount of ground water flowing into the cracks of the caves, the types of rocks found in the cave, the makeup of the ground water, and the humidity and temperature of the cave.

Some of the most common deposits found in caves are stalactites and stalagmites. These formations are created when water containing dissolved minerals (such as calcium carbonate) drips down from the ceiling of the cave. Stalactites hang downward from the ceiling while stalagmites grow upward from the floor of the cave.

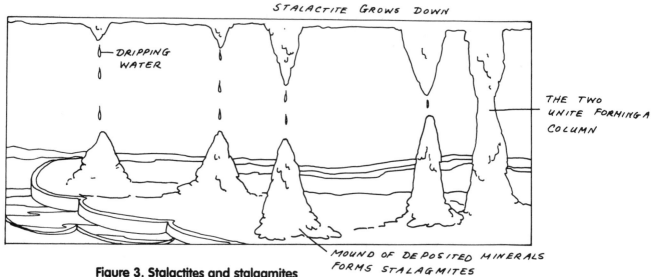

Figure 3. Stalactites and stalagmites

Pre-Lab Questions

1. How were limestone deposits formed? _____

_____

2. How do cracks occur in limestone rocks? _____

_____

3. How is carbonic acid formed?_____

_____

4. How does carbonic acid affect limestone?_____

_____

5. What mineral is usually found in water that forms stalactites and stalagmites? _____

_____

Name _____

A CAVE RAVE

I. Completion. As you tour the cave, complete the following. Record very specific answers.

1. Describe and draw the deposits you see inside the cave.

2. Are there more stalactites or stalagmites in the cave?_____

_____

3. Describe the texture (feel) and color of the cave deposits. _____

_____

4. What do you think would happen to a cave deposit if you were to pour some vinegar (a mild acid) on it? _____
What made you arrive at that conclusion? _____

_____

5. Can a cave change in size over time?_____ If you answered yes, explain how this is possible. _____

_____

6. How is the temperature and humidity inside the cave different from the temperature and humidity outside the cave? _____

7. Are there some structures in the cave that look like a stalactite and a stalagmite connected? _____ Explain how you think these are formed.

_____

_____

8. Is there any evidence of animal life within the cave? _____ If so, describe the evidences and what kind of animals you think may live there. _____

_____

9. Do plants live in caves? _____ Why or why not?_____

_____

_____

10. Imagine you are a cave-dwelling animal. In the space below, write a story about your life in the cave. Explain why you live there, what kind of animal you are, your daily routine, what you eat, and other interesting facts about your life.

ANSWER KEY

The Wonders of Wood
Pre-Lab Questions
page 4
1. Answers may vary but might include medicine, cosmetics, food, paper, and lumber.
2. Answers may vary but might include information about trees as producers of oxygen, conservers of soil and moisture, and habitats for animals.
3. Answers may vary but might include newspaper or cardboard.
4. Answers may vary but might be similar to the following story. Workers enter the forest in specially designed equipment. They reach around my underside and cut me off about two feet above ground level. I fall for the first time in my young life. The lumber equipment slides up my length, breaking off limbs and scattering pieces of bark. It loads me onto a big truck with my friends and relatives. We ride to the lumber yard, knowing our fate. When we arrive, we are dumped out onto the ground where we lie for several weeks. Eventually, we are cut with the big saw. After cutting, we no longer recognize ourselves because now we are a stack of lumber.

Answers in Data Collection, Ranking, and Thought Questions will vary.

Measurement Mania
Pre-Lab Questions
page 9
1. meter; gram (or kilogram)
2. Area can be calculated by multiplying length times width.
3. square meters
4. Volume can be calculated by multiplying length times width times height.
5. cubic meters

Conclusion Questions
page 11
1. Answers will vary.
2. You could have used your foot as a measurement tool. Calculate the length and width of the hall in (your) feet; then multiply that data by the known length of your foot.
3. Not everyone's foot was the same length as the king's.
4. So that when people all over the world discuss measurement, they are discussing the same quantities.

Machine Maneuvers
Pre-Lab Questions
page 15
1. lever, wheel and axle, pulley, inclined plane, wedge, and screw
2. to make work easier to do
3. It is easier to push something up an inclined plane than to lift it (inclined plane increases distance traveled, but reduces force required).

Answers in Data Tables I and II will vary.

Making Sense of a Situation
Pre-Lab Questions
page 21
1. to perceive their environment
2. sight
3. Light enters the eye through the transparent cornea. The amount of light entering the eye is controlled by the iris. As light passes through the lens, it is bent and focused on the retina at the back of the eye. These receptors are connected to nerves that send information about light perception to the brain.
4. The smell receptors are a patch of cells in the top of the nasal passage. There are about 5 million cells in this area that can detect certain chemicals in the gaseous state.
5. They both detect chemical information.
6. When objects touch the skin, they deform the sensory receptors in the skin, and this sends an impulse to the brain.
7. to the brain

Answers on student activity pages will vary.

More Power to the Games
Pre-Lab Questions
page 26
1. by multiplying m x a. m = your mass (weight) in kilograms
   a = acceleration due to gravity (10 m/sec^2)
2. 40 kilograms x 10 meter/sec/sec = 400 newtons
3. by multiplying force by distance through which the force moved
4. by dividing work by time required to do work
5. The child lifting the book did more work. Nothing was moved by the horse; therefore, no work was done.

Conclusion Questions
page 28
1. the 100-meter race. Work equals force multiplied by distance; the greater the distance, the more work done.
2. The three-second race. Power equals work divided by time; the less time required, the more power generated.
3. none; none
4. Answers will vary. Horsepower originally referred to how fast a horse could move a mass through a distance.

Journey to Yellowstone
Pre-Lab Questions
page 31
1. Wyoming
2. bison
3. geysers
4. Water and steam heated inside the earth reach the surface and spew out.
5. Answers will vary.
6. They become dependent on food given by humans.
Field Trip Questions
1. July, August
2. July, August
3. Old Faithful
4. brucellosis
5. buffalo
6. grizzly; they are carnivores (fish, etc.)
7. they are herbivores; eat plants
8. bighorn; great balance
9. Wolves can kill their herds of animals.
10. $20
11. Montana; any new bison entering specific areas in the park would be slaughtered and shipped out for testing.

Conclusion Questions
page 33
Answers will vary.

Treasures in the Forest
Pre-Lab Questions
page 37
1. The top of the forest, the canopy, is composed of the tallest trees with branches that overlap one another. The understory consists of trees that are shorter than the canopy trees. Shrubs are woody plants with many stems. They typically grow tall enough to reach the lowest branches of the understory trees. The herb layer is made up of short plants, such as the ferns that grow on the forest floor.
2. Answers may vary; sample answers are forest, field, and stream.
3. Answers may vary.
4. The forest provides different things to different organisms. It provides meals and homes to a variety of wildlife. Woodpeckers feast on the insects in trees, while redbirds build their nests in the tree branches. Trees provide these animals with shade, protect them from wind, act as noise and air pollution filters, and supply oxygen. The forest also provides valuable resources for humans. Pulp made from trees can be used to manufacture paper and wood products. Humans also enjoy many of the fruits produced by trees.

Answers on student activity pages will vary.

Acting Up
Pre-Lab Questions
page 42
1. Answers may vary; in warm, moist soil in the forest.
2. Organisms are born with instinctive, or inborn, behavior. Learned behavior is acquired through experience.
3. simple
4. It is displaying an instinctive behavior to make it look larger when it is alarmed.
5. decayed plant matter in the dirt
6. **Taxis** is movement in response to a stimulus. Negative taxis is movement away from a stimulus, while positive taxis is movement toward a stimulus.

Conclusion Questions
page 44
1. Answers will vary.
2. positive taxis with soil; negative taxis with ammonia
3. Their response helps assure they will avoid chemicals that could be dangerous.
4. Inborn. The worms did not learn these behaviors.
5. Answers will vary.

Classy Cars
Pre-Lab Questions
page 47
1. Answers will vary; bread is classified in a grocery store; shoes are classified in a department store.
2. When living things are classified, it is easier to study them, characterize them as individuals, and discuss them with people from other countries.
3. kingdoms
4. species
5. the genus and species name of an individual

Conclusion Questions
page 52
1. Answers may vary; cars are classified by style or type on car lots.
2. 16
3. Taxonomists classify things.

4. Answers will vary; in the lunchroom food is classified, in the main office mail is classified, in the attendance office student names are classified, in the storage closet cleaning materials are classified, and in the copy room types of paper are classified.
5. Answers will vary; students might classify music by types, such as country, rock, and classical.

Seed Hunt
Pre-Lab Questions   page 55
1. A **seed** is an embryo plant and the food needed for its early development.
2. A covered seed is inside fruit; a naked seed is not.
3. A **fruit** is a mature or ripened ovary.
4. wind, water, and animals
5. Answers may vary; pine and dandelion are two examples.

Digging in the Dirt
Pre-Lab Questions   page 60
1. Those in warm climates; because chemical reactions occur faster in warm climates than in cool and because there is a larger mass of plant life growing in warm climates.
2. **Topsoil** is the upper layer of soil; it is more fertile than subsoil because it contains humus.
3. Loose soil; plants and animals must have food, water, and air to survive; all of these are available in loose soil.

Conclusion Questions   page 63
1. Answers will vary; soil from pathways, roadways, or playgrounds is very compacted.
2. Answers will vary; soil in a forest or other areas where plants are living is often more moist than soil on pathways or playgrounds.
3. Answers will vary; soil from an area where people do not walk will contain more soil animals. Foot traffic compacts soil and reduces the air spaces in it, making it difficult for soil organisms to survive.
4. Soil in the woods; it contains a lot of dead plants and is not as compacted as soil on a pathway.

Sewage Safari
Pre-Lab Questions   page 67
1. a variety of sources, including sinks, toilets, and washing machines
2. It may contain bacteria that can cause disease.
3. a) primary—large materials are filtered from wastewater
   b) secondary—sludge is treated with bacteria to break it down into simpler substances
   c) tertiary—wastewater is treated with chemicals to kill bacteria
4. **Sludge** is the collection of heavy particles that settle out of wastewater in the large tank. It can be broken down by living organisms into simpler substances.
5. It is collected and burned as fuel.
Answers on the student activity pages will vary.

Aquarium Adventures
Pre-Lab Questions   page 72
1. Marine fish live in salty water; freshwater fish do not.
2. Answers may vary. Marine invertebrates: worms, sponges, coral. Marine vertebrates: fish, whales, sharks.
3. Examples from each group may vary. Mammal—whale, Birds—flamingo, Reptiles—alligator, Amphibians—frog, Fish—tuna
4. A streamlined body has a small head, a wider body, then a narrow tail. It helps the fish move through the water efficiently.
Answers on student activity pages will vary.

It's a Zoo Out There!
Pre-Lab Questions   page 79
1.

Mammals	Birds	Reptiles	Amphibians	Fish
Whale Platypus Flying Squirrel	Ostrich	Gila Monster	Toad Newt	Brim

2. Answers may vary. A student's habitat is his or her home.
3. Answers may vary. Polar bears are well equipped for the Arctic cold with thick coats of white fur on top of several layers of fat. The claws on its feet help it catch and kill prey, as well as swim. The white color serves as camouflage on ice. Fish have slender, streamlined bodies that move easily through the water. Their tail and fins propel and maneuver them. Their gills remove oxygen from the water.
Answers in the Data Table will vary.

A Mooooving Experience
Pre-Lab Questions   page 84
1. Farmers sat on stools, squeezed the cow's teats, and collected the milk in buckets.
2. An **udder** is where the cow makes and stores her milk. The udder is divided into four separate quarters. Each quarter has its own milk supply.

When the udder becomes full of milk, each section can be emptied through one teat. A **teat** is the projection through which milk is drawn from an udder.
3. Answers will vary; human hands can contaminate the milk if they are not washed well.
4. to feed their calves
5. A hormone is released that sends milk into ducts and then into teats. This hormone is produced when the calf brushes against the teat, or when the farmer washes it.
6. When cows are scared or upset, they will not produce milk.

Answers on the student activity page will vary, depending on the farm you visit. Some possible answers are listed below.
1. black and white; Holstein
2. produces greater quantity of milk
3. Answers will vary.
4. Washing stimulates milk to pass into teats.
5. storage tank; very cool
6. harmful bacteria
7. 24 to 48 hours
8. mastitis—blood in the milk from swelling and infection in the udders
9. Answers will vary.
10. goats
11. Answers will vary.

A Cave Rave
Pre-Lab Questions   page 88
1. from deposits of skeletons of sea animals
2. Earthquakes pushed up rocks and cracked them.
3. from the combination of carbon dioxide and water
4. breaks it down and slowly dissolves it over time
5. calcium carbonate
Answers on student activity pages will vary.